GREAT CHEMISTRY BOOKS:

A PERSONAL VIEW

Harold Goldwhite
Professor of Chemistry
California State University, Los Angeles

About the author:

Harold Goldwhite is Professor Emeritus of Chemistry, California State University, Los Angeles, where he has been on the faculty since 1962. He was educated in England (Cambridge University B.A. 1953; Ph.D. 1956) and was on the faculty of the University of Manchester Institute of Science and Technology from 1958 – 1962.

He co-developed a course on history of chemistry with the late Professor Anthony J. Moye and has been teaching it since the 1960s.

Dr. Goldwhite has authored or co-authored 9 books including "Chemical Chrestomathy", a series of brief biographies of chemists (available on-line); "Creations of Fire" (with Cathy Cobb), a one-volume history of chemistry; and "The Chemistry of

Alchemy " (with Monty Fetterolff and Cathy Cobb).

Dr. Goldwhite is married; has 4 children and 3 grandchildren; and lives in South Pasadena on the outskirts of Los Angeles.

ACKNOWLEDGEMENTS

I thank my wife Marie, as always; and my family, for their unfailing support and unstinting praise.
I thank my long-time friends Carola and Bruce Kaplan for suggesting I include a chapter on the book by Primo Levi.

PREFACE

I recently bought a fascinating book by Brian Clegg entitled "Scientifica Historica". (With a title like that I couldn't resist). The subtitle gives it away: "How the world's great science books chart the history of knowledge". Since imitation is the sincerest form of flattery I am writing this parallel exploration of a personal selection of the world's great chemistry books. (Clegg's work, covering all the sciences and mathematics, is a bit thin on chemistry). The period I cover is from antiquity through the 20th. Century. The title will be taken seriously; only books will be included. There are papers and pamphlets that I may mention in passing, but books are the theme. Of course my choices are idiosyncratic and you may vigorously disagree.

These chapters appeared initially, in slightly different form, as columns in two publications: SCALACS, the journal of the Southern California Section of the American Chemical Society; and The Indicator, the journal of the New York and New Jersey Section of the American Chemical Society. I am grateful to the editors of these journals for their long-term support of my columns about the history of chemistry.

CONTENTS

CHAPTER 1   Plato; and Lucretius

I start with Plato because his dialog "The Timaos" was the only Platonic book, in dialog form, known for centuries in early and medieval Europe. The science historian J.R.Partington calls this dialog "perhaps the first extant treatise on chemistry". While Plato credits earlier philosophers with various element theories it is in Timaos that he firmly asserts the existence of four elements: earth, air, fire, and water. It is "out of four things of this kind [that] the body of the universe was created". I place such emphasis on the four element theory because, with relatively minor variations, it held sway over the minds and works of alchemists and chemists for some 2000 years. Indeed in the late

17th. Century Robert Boyle decided it was worth his effort to attack the four element theory in his book "The Skeptical Chymist".

Plato was apparently not an experimental scientist. He mistrusted, like many later philosophers, (Descartes comes to mind), knowledge gained solely from observation and experiment. Knowledge had to fit into a philosophical overarching scheme. Plato devised such a scheme in which 5 "Platonic" solids, the faces of which were regular geometric figures like equilateral triangles or squares, represented the four elements, and the foundation of the universe. An illustration: the tetrahedron, with its four triangular faces and sharp points, represents the penetrating element fire. The triangles or squares that are the basis of an element can unravel and form into another shape.

Consequently the transmutation of one kind of matter into another is quite possible As Partington says
" alchemy in germ goes back to Plato".

Plato and Aristotle – as imagined by Raphael

Perhaps my choice of Plato as the author of my first great book of chemistry was not a surprise. But my second choice may be. It is a narrative book-length Latin poem written by Titus Lucretius Carus, hereafter abbreviated to Lucretius (ca. 100 BCE to 55 BCE) who wrote "De Rerum Natura" in about 57 BCE. In this work Lucretius incorporates many ideas that come from the Greek philosopher Epikouros (Epicurus in Latin) who lived about 340 BCE to 270 BCE and who himself drew from the ideas of Demokritos (Democritus) who, alas, did not leave us a book. And neither did Epikouros. So at third hand, via Lucretius, we get our understanding of the atomic theories of these early Greek philosophers. It is worth noting that Aristotle criticized at length and dismissed the atomic theory of Demokritos on grounds, among others, that it included a void – a concept firmly rejected by Aristotle.

Demokritos – as imagined by a Roman sculptor

Nothing is ever begotten out of nothing by divine power" and the workings of the world take place without intervention by the gods. This is a radical idea, foreshadowed by some of the earliest Greek philosophers, and quite opposite to conventional thinking of the period. "Nothing is ever annihilated, but all things on their dissolution go back into [the elements]". Void exists! The elements are made of "atoms" that are in ceaseless motion. (Lucretius did not use Demokritos' term; he speaks of seeds and first beginnings). Matter and space are infinite. Atoms differ in size, shape, and weight. There is a limit on the number of their shapes. Sense impressions like color, taste, smell etc. are not properties of atoms but are the effects of numbers of atoms on the percipient. Atoms may collide and become attached and thus, by accretion, things are created. Lucretius' poem explains at length how many natural phenomena can be explained on the basis of this atomic theory.

A 16TH. Century edition of Lucretius' poem.

Because of the widespread acceptance of Aristotle's ideas on pretty much everything from ethics to logic; from politics to poetry; from physics to biology, and because even his views on the creation of the universe were, in some sense, compatible with Christian doctrine, Aristotle became the major figure in philosophy in the early and medieval Christian church. Hence his views on the void, the constitution of matter, and his rejection of atomism became the orthodox doctrine in Europe until the 17th. Century. (See Robert Boyle mentioned above and discussed below).

Plato and Lucretius – Greek and Roman – two pillars of the ancient European world. Their speculations laid the basis of both theoretical and experimental alchemy and chemistry for millenia. These are truly great books of the precursors to chemistry.

CHAPTER 2. Geber (Jabir) "The Works of Geber".

Although the inclusion of an alchemical work in this exploration might be questioned, I believe, as do most historians of chemistry, that alchemy is a direct precursor of modern chemistry. And so such an exploration would be incomplete without a look at a representative work of alchemy –if there is such a thing! My choice, in part because I have a copy in my personal library, is "The Works of Geber" Englished by Richard Russell, 1678" in an edition by E. J. Holmyard, a distinguished historian of chemistry. My copy was published in 1928. Geber is the westernized version of Jabir, one of the most famous Arab alchemists. There is a large body of work ascribed to Jabir, who lived from about 720 C.E. to about 800 C.E. He was born in what is now Iraq and studied religion and mysticism – possibly including alchemy -in Arabia. He also became a proficient physician and served in that capacity at the court of Haroun-Al-Rashid, the caliph of the Arabian Nights. Because Jabir became a famous alchemist many later workers used his name to add luster to their own efforts. Consequently it is quite unclear how much of the extensive Jabirian corpus can be ascribed to Jabir himself. Scholars of alchemy have agreed that the core of Jabir's alchemy includes the following points.

Jabir's connection with world health.

While accepting Plato and Aristotle's four elements: earth, air, fire, and water, Jabir had a novel "insight" into the constitution of metals. They are made up of two "exhalations", one earthy and one watery. These are respectively the "sulfur" and the "mercury" principles. These principles are not to be equated to the familiar substances that bear these names. The names are a convenient shorthand for the essential essences that go into making up the familiar terrestrial substances. To turn to more practical aspects of Jabir's work it is clear that he was an experimentalist. He describes the operations of alchemy and the materials on which those operations were performed. He wrote a whole "book" (many of the "books" of this period are what we would call chapters) on furnaces. In Russel's English version there are illustrations of each type of furnace: calcinatory; distillatory; descensory; fusory; solutory; fixatory; and the water bath. However the writings of Jabir are often deliberately obscure. Alchemy was a "mystery" (or mastery) only to be passed down and interpreted by those whose lives were given to its study. These adepts would have the key to unlock a passage like the following (I have deliberately chosen a short extract): "Of the Calcination of middle minerals. All Atraments, Salts, Allones, and the kinds of Tutia are calcined in the said Calcinatory Furnace, with Tartar and other Things, with fire moderate or strong, according to the Exigency of Things to be calcined; as is evident in Our Book, "Of the Investigation of the Perfect Magistery"; but all bodies are calcined, as in Our Testament."

I promise you more clarity in my next chapter.

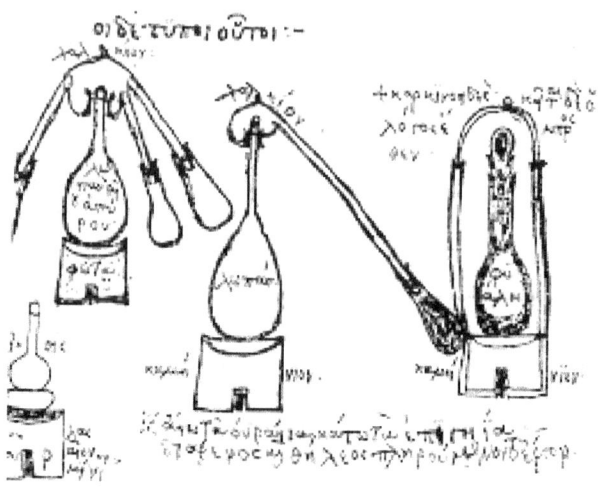

Alchemical equipment (from Berthollet's work.)

CHAPTER 3   Probierbuchlein;
Biringuccio, "Pirotechnia";   and
Agricola, "De Re Metallica".

I now turn from the mystic and almost undecipherable manuscripts of the great alchemist to the completely practical publications (yes; we move now to the era of printed books) deriving from the mining and metallurgical industries.

Growth in population in early medieval times led to increased demand for manufactured goods and a corresponding growth in extraction of mineral resources. This is reflected in a number of publications and I begin with the anonymous "Probierbuchlein" (pamphlets on testing) of the early 16th. Century. These were published as books in Germany, an area in which mining flourished, and they described, among other procedures, methods of separating gold and silver (a procedure called parting) since silver may be a contaminant of native gold.

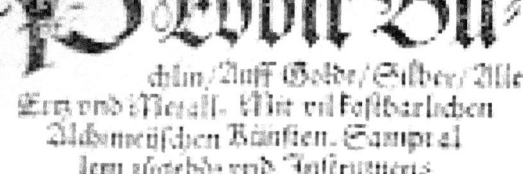

chlin/Auff Golde/Silber/Alle
Ertz vnd Metall. Mit vil kostbarlichen
Alchemeüschen Künsten. Sampt al
lem zügehör vnd Instrumen=
ten darzü dienlich.

⬥ Die BergEnamen/für die newen/
angohnde Bergkleüt.

⬥ Register füch am end.

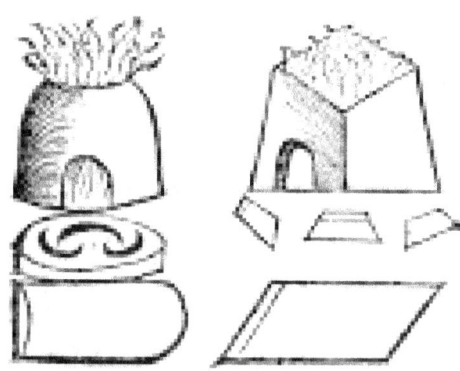

Zu Straßburg, Bei Christian Egenolph.

In 1540 Vanoccio Biringuccio (1480 – 1538) published "De La Pirotechnia" (on the technological applications of fire) a richly detailed and well-illustrated manual on metallurgy, applied chemistry, gunpowder, fireworks, and related subjects. Biringuccio had traveled in both Italy and Germany and had served as an advisor to many Italian nobles and the Pope. While the book is quite practical in its descriptions there are some discussions of theory. Biringuccio does support the four element theory, but is dismissive of transmutation. He notes that lead increases in weight by fire by about 8% (in its conversion to the calx, or oxide) and draws a parallel to animal corpses ("which weigh more than when alive"!)

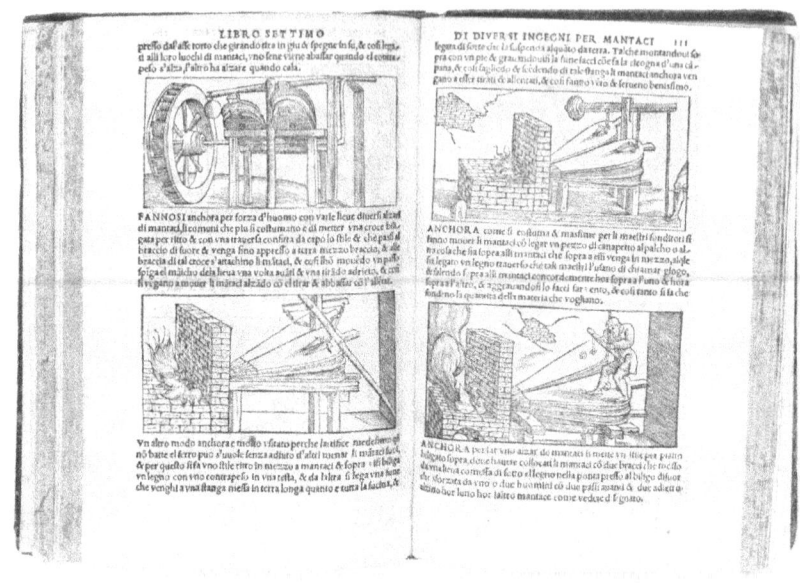

Illustrations from "Pirotechnia".

A decade later, in 1556, Georg Bauer (1494 – 1555), known by the Latinized name of Agricola, had his great book "De Re Metallica " published posthumously.

. Agricola was a prolific author and had produced an earlier work on mining and metals. "De Re

Metallica", with its many and very attractive illustrations, is his most important book. Some of Agricola's book is clearly copied from Biringuccio but much of it is original. Agricola, who was also a physician, lived in mining districts in Bohemia and learned much by direct observation. He is neutral on the topic of alchemy, and is clearly influenced by the writings of Paracelsus. He covers most of the topics that are included in Pirotechnia, but adds a section on glass-making taken from his own observations during an extended stay in Venice, then (and still) a renowned center for the craft.

\.

Agricola

Reasonably priced reprints of English versions of both Biringuccio and Agricola's major works are available.

The English translation of De Re Metallica was made by two authors, one of whom became the President of the United States. I'll leave it to my readers to work that one out.

CHAPTER 4. Paracelsus; "Archidoxis"

You may wonder why I include a workby such a strange character as Paracelsus in my survey of great chemistry books. After all he was an itinerant physician and alchemist, intolerant of the conventional medicine of his time, and constantly embroiled in controversy.(For a more detailed biographical sketch of Paracelsus see "Creations of Fire" by Cobb and Goldwhite). He wrote (or perhaps dictated) over a hundred books that were unpublished in his lifetime but were assiduously collected, edited, and published by his disciples. The seven volumes of "Archidoxis" were probably written around 1526 but first published (in Krakow) in 1569. (I confess that I have not seen this book and I am relying on the comprehensive discussion in Volume II of J.R. Partington"s "History of Chemistry).

Paracelsus' work is worthy of inclusion as a great book because it altered the emphasis of alchemy/chymistry in a way that changed it into a discipline that we can recognize today. Following precursors including John of Rupescissa and van Helmont, Paracelsus insisted that the goal of his work was to use alchemy to find cures for diseases. He coined a term to describe his new direction: he called himself an iatrochemist – and many of his followers adopted this new coinage.

A contemporary image of Paracelsus.

Alchemy should not be about making gold or silver (though Paracelsus was an alchemist and believed in the possibility of transmutation) but about exploring both mineral and other sources to develop medications. The "Archidoxis" is important not for its collection of alchemical recipes and procedures, though these are unusually clear and complete in contrast with most alchemical works, but for its philosophy. Along with other Paracelsian writing it changed the direction of chemistry into a path still followed today.

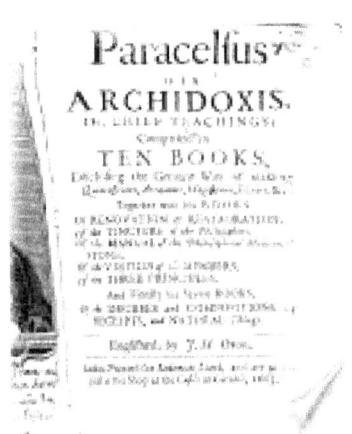

CHAPTER 5: Libavius; "Alchemia":

The first publication that can really be called a textbook of chemistry is "Alchemia" by Libavius.

Libavius (Libau), was born in Saxony in 1540; and died in Coburg in 1616. He was the son of a weaver. He gained entrance to the university at Jcna where he earned an M.D. He became a teacher, including a stint as Professor of History and Poetry (!) at Jena. For the rest of his career he alternated between medicine and teaching, developing an interest in chemistry. In his medical practice he used chemical medications including, following Paracelsus, potable gold. He gave credence to transmutation in his writings, but warned against fraudulent practices by pseudo-alchemists. The general tone of his writings, which were voluminous, is that of a careful academic.

"Alchemia" was published in 1597 at Frankfurt and is in Latin, the scholarly language at that time in Western Europe. It is a very long large-format lavishly illustrated book that must have been very expensive. Oddly enough it was printed on rather poor paper, and in several parts, some of which are now extremely rare. Part I has plans for Libavius' ideal chemical house that includes laboratories, a museum, a wine cellar (no doubt conducive to chemical thinking), and gardens.

As Partington, in Volume 2 of his "History of chemistry" states: "Libavius' Alchemia is an excellent practical text-book … a clear, concise and sensible style, entirely different from the rambling, bombastic, and obscure verbiosity of Paracelsus or the alchemical authors." Alchemy is carefully defined and generally supported, but without overly enthusiastic claims.

"Alchemia" includes some initial attempts at qualitative analysis especially of mineral waters. At the time they were widely used in medicine. Libavius shows that most natural mineral waters contain a variety of dissolved minerals including, depending on the source, common salt, nitre, vitriol, and alum. He does not distinguish between inorganic and organic compounds. There is a clear description of the preparation of nearly pure ethanol, vinum ardens, from wine or beer. When litharge (lead oxide) is dissolved in vinegar the crystalline sugar of lead is produced , so-called because lead acetate allegedly tastes sweet. (Safety note to readers: do NOT try this!) Heating lead acetate yields a "quintessence" – volatile and flammable (acetone).

Complete original copies of this foundational textbook, "Alchemia", are rare. I have never seen or held one, though I do have a facsimile reprint. Authoritative sources denote it as the first real chemical text and so it finds its place among the Great ChemistryBooks.

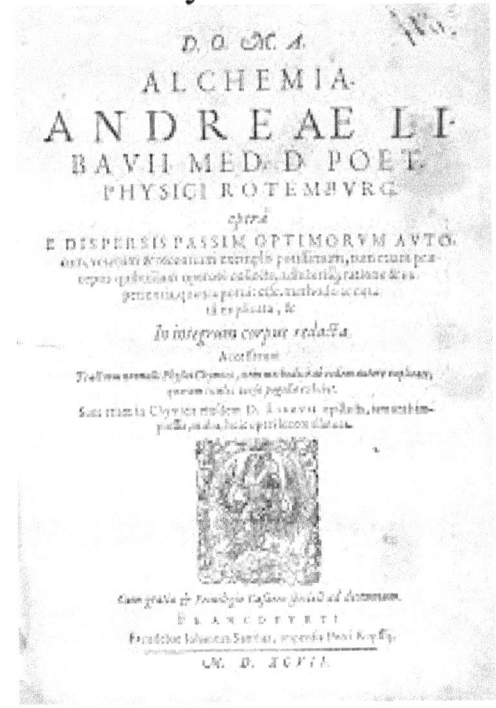

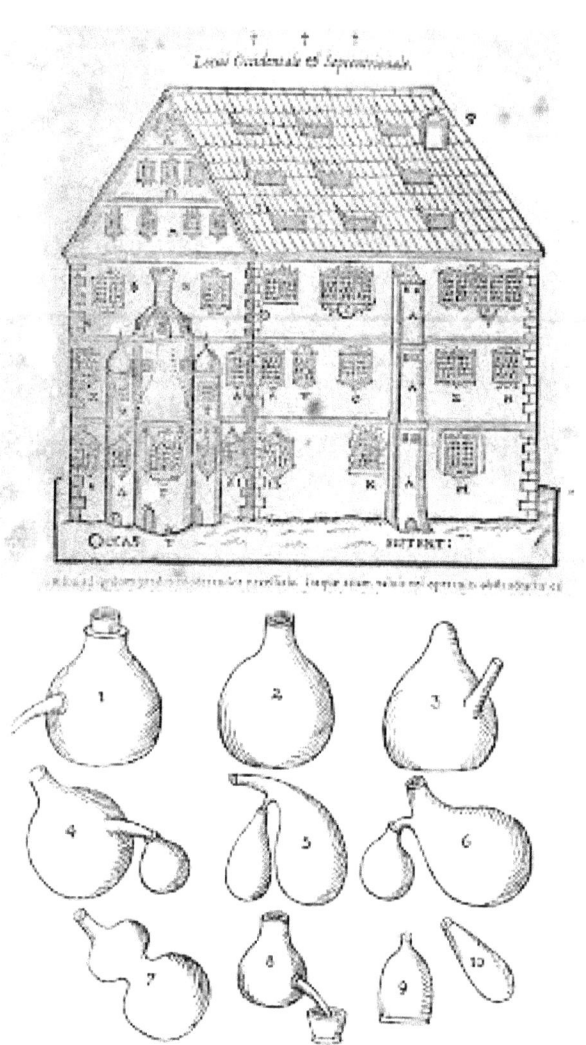

Above: Ideal chemistry house
Below: Glass vessels

From "Alchemia"

CHAPTER 6:   Robert Boyle; "The Sceptical Chymist"

I come now to the 17th. Century and to that remarkable work by the Honorable Robert Boyle, "The Sceptical Chymist". The first English Edition appeared in 1661 and there were many subsequent editions in both English and Latin. My own modest copy is one of Dent and Dutton's Everyman's Library and was Edited and with a Foreword by M.M. Pattison Muir, a distinguished historian of chemistry, whose book "A History of Chemical Theories and Laws", published in 1907, is still excellent reading.

Boyle was one of the greatest natural philosophers of the 17th. Century. His study of gases (airs) is recalled in Boyle's Law. He worked on acids, bases, and indicators; published the first clear description of the preparation of elemental phosphorus ("the icy noctiluca"!); was a close friend of Isaac Newton; and, like Newton, was an alchemist.

The Sceptical Chymist is not an easy read. It is written in the then popular form of a dialog between Themistius, a follower of the ancient beliefs of the hermetick philosophers (alchemists), and Carneades, a thinly disguised Boyle. The hermeticks believed in the four elements (earth, air, fire, and water) as the bases of all material things. Carneades challenges these ideas, insisting on experiments, verified facts, and reasoned beliefs – all notions going back to views expressed nearly a century earlier in Francis Bacon's "Novum Organon", a work that enunciates clearly, and perhaps for the first time, what we now call the scientific method.

In attacking the ancients' views of elements Boyle writes "I mean by elements, as those Chymists that speak plainest do by their principles, certain primitive and simple, or perfectly unmingled bodies; which not being made of any other bodies, or of one another, are the ingredients of which all those called perfectly mixt bodies are immediately compounded, and into which they are ultimately resolved; now whether there be any one such body to be met with in all, and each, of those that are said to be elemented bodies, is the thing I now question."

Boyle failed to resolve the experimental question as to whether a particular substance was an element. Indeed he argued that not a single unambiguous example of an element was known to him. This book is more of an exercise in destruction than construction. Boyle argues powerfully against the four elements of the hermeticks. And he provided a definition of a true element that has a modern and convincing sound. But it took another century before Lavoisier, in another book that I plan to discuss later, provided the true quantitative test of whether a substance was indeed an element.

I conclude with a quotation from Pattison Muir: "The great importance of the Sceptical Chymist consists in Boyle's reiteration of proofs that nature is not simple, but rather overpoweringly complex; of proofs that it is wise to doubt every short and easy road to natural truths;….that above all "occult qualities" are nothing but "the sanctuary of ignorance." Words to live by.

The Sceptical Chymist

Robert Boyle

THE

SCEPTICAL CHYMIST:

OR

CHYMICO-PHYSICAL

Doubts & Paradoxes,

Touching the

SPAGYRIST'S PRINCIPLES

Commonly call'd

HYPOSTATICAL,

As they are wont to be Propos'd and
Defended by the Generality of

ALCHYMISTS.

Whereunto is præmis'd Part of another Discourse
relating to the same Subject.

BY

The Honourable *ROBERT BOYLE*, Esq;

LONDON,
Printed by *J. Cadwell* for *J. Crooke*, and are to be
Sold at the *Ship* in *St. Paul's* Church-Yard.
M DC L I.

CHAPTER 7: Leméry: "Cours der Chimie"

Most of the works I have discussed or will discuss in upcoming chapters dedicated to "Great Books of Chemistry" are by authors whose names are likely to be familiar to you. But this chapter's book is by someone you probably have not heard of unless you are quite familiar with 17th. Century science. This author is Nicolas Leméry, who was born in Rouen in November 1644 or 1645, and who died in Paris in 1715. His father was a lawyer to the Normandy Parliament and a Protestant. Nicolas was apprenticed to a pharmacist who was a relative, but he moved to Paris in 1666 to work at the Jardin du Roi, the Royal Botanic Garden that included science laboratories, and sponsored science lectures. (Lavoisier, a century later, learned much of his chemistry at this institution.) Lemery's mentor was Glaser but they did not get on and he moved to Montpellier where he set up a laboratory, made and sold chemicals and pharmaceuticals, and began to give lectures on chemistry illustrated by experiments. These lectures established his fame and attracted many (paying) attendees including women and foreign students.

Lemery's Protestant faith led to conflicts with the Catholic authorities and from 1683-84 he visited Protestant England. Returning to France he earned an M.D. and he became a practicing Catholic in 1686 and was accepted into the Paris Academy in 1699.

In 1675 Leméry published his "Cours de Chymie" which became by far the most popular chemistry text published to that date and is the reason for the inclusion of Leméry in this book. The "Course of Chemistry, containing the way of carrying out the operations useful in medicine, with the rationale for each operation, for the instruction of those who want to apply themselves to this science" was a volume of over 500 pages with many illustrations. It went through at least 11 authorized editions in French in the next 40 years; there were also pirated editions. The work was first published in Enghlish in 1677 and three further editions appeared, the last in 1720. The "Cours" was also translated into German, Dutch, Italian, and Spanish.

Leméry also compiled a popular Pharmacopeia (first edition in 1698 and a monograph on antimony (1707) that echoes Basil Valentine's "Triumphal Chariot of Antimony".

Leméry's text is clear and reasonably concise. It follows on from earlier German and French authors, particularly Le Fèvre, and shows Paracelsian influences in its endorsement of remedies derived from antimony and other metals. Its period is before the rise of the phlogiston theory and it was eventually replaced by texts that embraced the newer ideas. Leméry was unimpressed by the claims of alchemy. He described them as mostly trickery. However he does support an element theory; his five elements or principles are mercury, oil, sulfur, water and earth. The incorporation of the fire principle into sulfur is familiar in the writings of many Arabic alchemists.

The "Cours" is eminently practical and describes in detail the apparatus, including furnaces, needed to embark on practical chemistry. It includes a glossary and tables of symbols. The body of the book covers the three separate areas of minerals; vegetable materials; and animal materials. These three divisions were not novel to Leméry but his book helped popularize them. He did add one new theoretical speculation derived from the ideas of the corpuscularists (including Gassendi, Descartes, Boyle and Newton) – that corpuscles might have particularly shaped spikes and orifices. This became a popular way of explaining acid/base interactions.

A clear, comprehensive, and popular view of practical chemistry in the late 17th. Century Nicolas Leméry's "Cours de Chimie" takes its place among the "Great Chemistry Books".

Title page of "Cours de Chimie" with Leméry's portrait.
Next page: Image of a Laboratory from the "Cours".

CHAPTER 8: Joseph Priestley: "Experiments and Observations on Different Kinds of Air"

The eighteenth century saw great advances in chemistry, sparked by the rapid growth in studies of gases, or "airs" as they were called at the time. The initial spark came from seventeenth century observations by Van Helmont, who coined the term gas from a Greek root meaning chaos, and who realized that there were gases different from common air, which was still regarded by many natural philosophers as one of the four elements. The following century saw great advances in pneumatick chemistry, as the study of gases was called. The great chemistry book that I will consider in this column incorporates much of the research undertaken by a seminal figure of pneumatick chemistry. The researcher is Joseph Priestley (1733 – 1804); the book is "Experiments and Observations on Different Kinds of Air" (3 volumes, 1774 – 1777).

Priestley was an impressively talented man, by profession a dissenting minister, i.e. a Christian but not an adherent to the established Anglican Church; and by enthusiasm a natural philosopher. He wrote and published prodigiously: charts of history and biography; a history of electricity; a history of vision, light, and colors; several versions of the work on airs already mentioned; two books attempting to establish the doctrine of phlogiston; a book on oratory (he had a stammer!); and around "50 works on theology, thirteen on education and history; eighteen on political, social, and metaphysical subjects; and twelve books and about fifty papers on scientific subjects" (Partington, "History of Chemistry", Vol.3).

Given such productivity it is not surprising to learn that the Editor of "Philosophical Transactions", the journal of The Royal Society, after receiving a number of voluminous papers on airs from Priestley, intimated to him that there was no room in Phil. Trans. for any more papers by Priestley, and that he should consider publishing elsewhere. Thus the "Experimental Observations...".

Several chapters could be devoted to the career of Joseph Priestley, but those are for another place. For now I will focus on Priestley's numerous discoveries and observations in pneumatick chemistry. The pneumatick trough, invented by Stephen Hales, separated the generator and the collector of airs, and was an advance in technique. Priestley used two types of pneumatick troughs, collecting airs over water or over mercury. In this way he was able to isolate airs soluble in water.

Early in his work, when he lived in Leeds next to a brewery, he began studying mephitic air (carbon dioxide) produced in the fermentation of grains to beer. He found that impregnating water with mephitic air gave a refreshing sparkling beverage similar to natural mineral waters. Priestley recommended his new beverage to the Royal Navy as a possible cure for scurvy (it wasn't). His pamphlet on this discovery was his most popular publication, and led to an industry that flourishes to this day – the carbonated beverage industry.

The airs Priestley discusses in his book include nitrous air (nitric oxide); phlogisticated air (nitrogen); nitrous vapor (nitrogen dioxide); nitrous air diminished (nitrous oxide); acid air (hydrogen chloride); inflammable air (hydrogen); vitriolic acid air (sulfur dioxide); fluor acid air (silicon tetrafluoride); alkaline air (ammonia); carbon monoxide; and, most significant of all, dephlogisticated air (oxygen). Priestley was not always the earliest discoverer of some of these airs, but he excelled in the wide scope of his work and in his extensive examination of their properties, particularly those of oxygen. Priestley's work on oxygen was the key to Lavoisier's revolution in chemistry and the improved understanding of combustion and respiration. With Priestley's great book on airs we are on the threshold of a new era of chemistry which is encapsulated in the next great book I plan to discuss.

Priestley; and a typical experiment involving a pneumatick trough.

CHAPTER 9: Antoine Laurent Lavoisier : " Elementary Treatise on Chemistry".

The previous chapter was on the great contributions that Joseph Priestley made to pneumatick chemistry, many of which were recorded in his great book on "Experiments and Observations on Different Kinds of Air" (3 volumes, 1774 – 1777). In this chapter I will discuss the revolutionary work on chemistry published by Antoine Laurent Lavoisier (1743 – 1794) in 1789 entitled "Elementary Treatise on Chemistry". This book was translated into English in 1790 by the Scottish chemist Robert Kerr and retitled "Elements of Chemistry'. In 1965 Dover Books published an inexpensive facsimile copy of Kerr's version that was in print for many years.

The subtitle of "Elements of Chemistry" is "in a new systematic order, containing all the modern discoveries". The introduction to the Dover reprint is by Douglas McKie, a major historian of science and author of a superb one volume biography of Lavoisier. This 30-page introduction gives a summary of Lavoisier's career and of the events that led up to Lavoisier's revolution in chemistry. That revolution was the proposal that the phlogiston theory of combustion must be replaced by the oxygen theory. Oxygen, discovered independently by Priestley and Scheele and named by the former dephlogisticated air and by the latter fire air, was renamed oxygen by Lavoisier (meaning acid-former) and was presented by Lavoisier as the key to combustion and respiration.

Throughout "Elements" Lavoisier sticks rigorously to quantitative methods, and includes mass balances, and by implication the law of conservation of mass, in his calculations. The new systematic order promised in the subtitle begins with a first part "On the Formation and Decomposition of Aeriform Fluids – of the Combustion of Simple Bodies, and the Formation of Acids". This section ends with Lavoisier's "Table of Simple Substances", i.e. elements in modern nomenclature, an innovation. While this Table starts off with Light and Caloric (heat) it follows with oxygen, azote (nitrogen) and hydrogen, and then describes the non-metals (Oxydable and Acidifiable simple Substances not Metallic) including Sulphur, Phosphorus, Charcoal (carbon), and 3 still unknown radicals: muriatic (chlorine), fluoric (fluorine), and boracic (boron). Throughout the text Lavoisier uses the revolutionary new nomenclature of chemistry that he, with his colleagues Berthollet, Fourcroy, and De Morveau, had developed and published a few years earlier.

Part II of the text is "On the Combinations of Acids with Salifiable Bases, and of the Formation of Neutral Salts" and includes not only Lavoisier's own work but also most of the work in this area reported by others.

Part III is a "Description of the Instruments and Operations of Chemistry"; a comprehensive introduction that includes gas handling and weighing; thermometry; calorimetry; distillation; fermentation (!); and the operation of furnaces. At the back of the text is a series of Plates, many based upon sketches made by Madame Lavoisier who was a skilled artist and Lavoisier's note-taker, secretary, and companion in much of his laboratory work.

The Appendices include many conversion tables of measurements of length, mass, and temperature from French into common English units – including the degrees of Reaumur's Thermometer into its corresponding degrees of Fahrenheit's Scale. These tables remind one of the important work that Lavoisier did as a member of the Commission set up after the French Revolution to produce uniform systems of measurement in France – the beginning of the metric system. Despite this work and all the important advances in science contributed by Lavoisier he was condemned during the Terror of offences against the People by being a member of a tax-collecting syndicate and was guillotined at the age of 51.

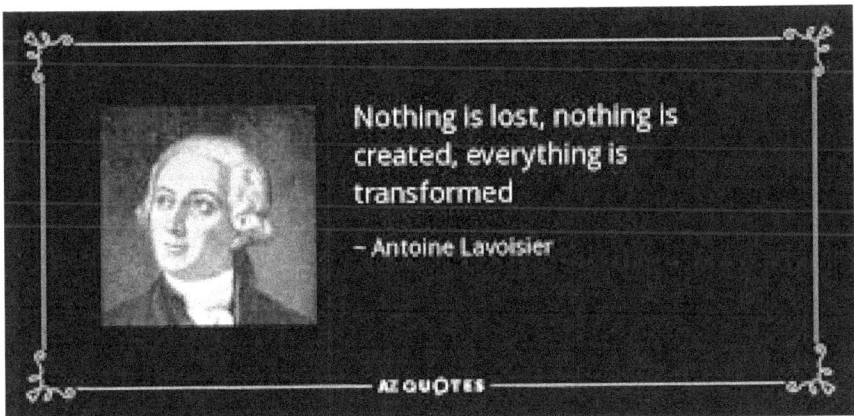

Nothing is lost, nothing is created, everything is transformed

– Antoine Lavoisier

AZ QUOTES

Lavoisier and Madame Lavoisier: an experiment on human respiration.

CHAPTER 10: Guyton de Morveau, Antoine Lavoisier, Claude-Louis Berthollet, and Antoine-Francois Fourcroy. "Method of Chemical Nomenclature".

In the previous Chapter I made a passing reference to a book on nomenclature. Upon reflection I have decided that the book I referred to in passing is, in fact, well worthy of inclusion in my list. It is the multi-authored "Méthode de Nomenclature Chimique" by Guyton de Morveau, Antoine Lavoisier, Claude-Louis Berthollet, and Antoine-Francois Fourcroy, published in 1787. The authors are listed in order of seniority by age. Each of them contributed to the text, but the first listed author,

Guyton, was the leading spirit of the enterprise.

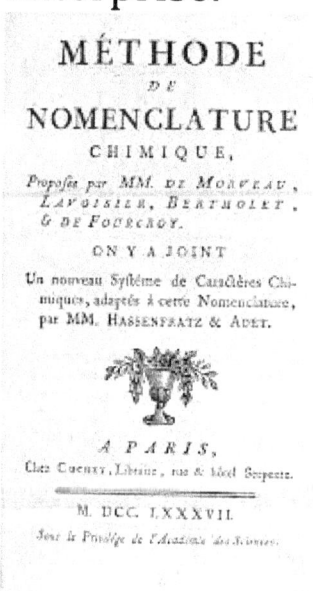

Nomenclature may be considered by many as a rather dull subject. But no less an authority than M.M. Pattison Muir, a distinguished historian of chemistry, wrote in his "A History of Chemical Theories and Laws" (1907): "to write a full description of the origin, growth and misadventures of the language of chemistry is to write a history of the science." Full disclosure: I copied this quotation and its source from a marvelous historical work by Maurice P. Crosland titled "Historical Studies in the Language of Chemistry" published originally in 1962 and reprinted by Dover Books, with a new Preface, in 1978. It is out-of-print but worth pursuing in the on-line catalogs of used-book sales sources. Much of this chapter is inspired by Crosland's book and also Volume III of J.R. Partington's "A History of Chemistry".

To understand the significance of the Method (as I shall refer to it) we must look back to the language of chemistry before its publication, and by that I mean the language of alchemy and chymistry, precursors to the modern science of chemistry. That language abounds with names that refer to incidental properties of elements and compounds:

quicksilver for mercury; oil of vitriol for sulfuric acid; sugar of lead for lead acetate; liver of antimony for antimony sulfide; – and the list goes on and on.

One impetus for the reform of chemical nomenclature was probably the reform of botanical nomenclature initiated by Carl Linnaeus in 1758. He introduced and elaborated the now familiar binary Latin names for plants that had previously borne common names not unlike those used by alchemists for chemical materials. The common garden pea becomes Pisum sativum; the ordinary beet is Beta vulgaris. This form of plant nomenclature is in use to this day. Torbern Bergman, a Swedish chemist who knew Linnaeus, was the first to see merit in applying a similar scheme to chemical compounds. He began to call fixed air (carbon dioxide) acidum aeratum (acid air) to describe a chemical property. But his approach lacked consistency.

And so we come to the Method. Inspired in part by earlier articles by De Morveau it boldly proposes a complete reform of chemical nomenclature, particularly of inorganic compounds, introducing (in French, of course) the binary naming of compounds that we use to this day. Oil of vitriol becomes Acide sulfurique; Acide sulfureux contains less oxygen. Salts formed from the former are Sulfate; the latter Sulfite etc. About a third of the Method is a comprehensive dictionary showing the equivalence of the new terms to those commonly in use. The new approach was presented to the Academie des Sciences in a lecture delivered by Lavoisier in April 1787 titled "On the necessity of reforming and perfecting the nomenclature of chemistry".

As is usual with any revolutionary idea the reform of nomenclature was not enthusiastically endorsed by many chemists. Resistance in the German speaking States, still holdouts for phlogiston, was especially strong. An interesting comment is contained in a letter written from Paris by Thomas Jefferson in July 1788: "One single experiment may destroy the whole filiation of his terms and his string of sulfates, sulfites...". Jefferson concluded that the reform was premature. Nevertheless as the oxygen theory began to prevail over phlogiston the new nomenclature began to prevail over the old. And with the introduction of a new symbolism by Berzelius in the early 19th. Century the pathway to the modern expression of chemical names and formulas was beginning to open.

CHAPTER 11: Berthollet: "Essai de Statique Chimique".

The author of the great book of chemistry in this chapter is Claude Louis Berthollet (1748 – 1822). The book is his "Essai de Statique Chimique" published in 1803.

Berthollet studied medicine in Turin, where he became M.D. in 1768. After successfully practicing as a physician he moved to Paris and studied chemistry with Macquer and Bucquet and earned a second M.D. in 1778 with a dissertation on the digestive properties of wines! As one of the group of rising young chemists in France he became a co-author with Lavoisier, Fourcroy, and Guyton de Morveau of their revolutionary book that began to reform chemical nomenclature (Chapter 9). He survived the reign of terror and became director of the Gobelin tapestry works where he introduced chlorine as a bleaching agent for textiles.

In 1794 Berthollet became Professor of Chemistry at the new École Polytechnique, an institution he helped to establish. Apparently his teaching was not appreciated by his students, and it was perhaps timely that he was chosen by Napoleon to be part of the team of scientists, archaeologists, and historians that the Emperor appointed to accompany the campaign in Egypt. This was the start of modern Egyptology. Among the great discoveries made by the expedition was the Rosetta Stone that allowed decipherment of hieroglyphics.

Among the great chemical discoveries made by Berthollet in Egypt were the first inklings of the Law of Mass Action. Berthollet, like many other chemists of his time, was profoundly interested in chemical affinity, that is the nature of the forces that attract some compounds so strongly that they react, while others remain inert. While exploring the Nile Valley he came across Lake Natron and made a startling observation. This brine lake, which was bordered by limestone cliffs, had deposits of sodium carbonate on its shores. Now Berthollet knew that in the laboratory by adding a solution of sodium carbonate to one of calcium chloride he had observed a precipitate of calcium carbonate (limestone) and a solution of sodium chloride (brine). He had "a flash of insight". Chemical affinity was not absolute; it could be affected by conditions of concentration and perhaps temperature. What he had observed at the lake was an example of a reversible or an equilibrium system. The actual composition of such a system depended on circumstances.

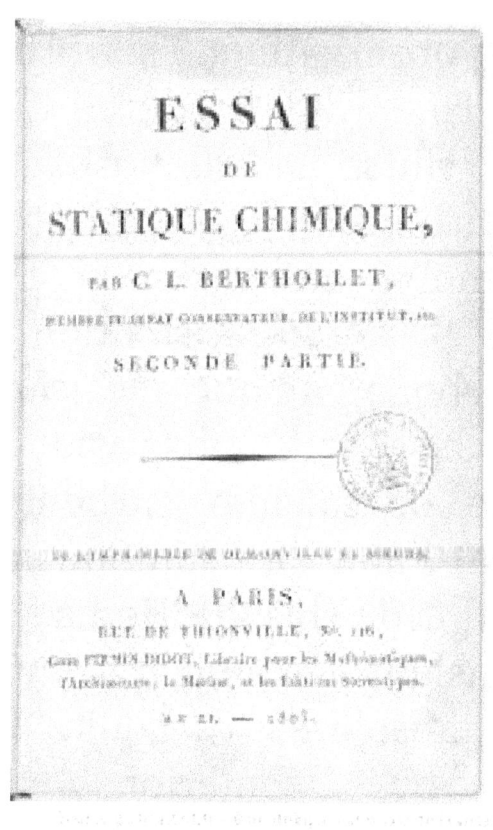

In the "Essay of Chemical Statics", to give it an English title, Berthollet took this idea and ran with it. He went so far as to challenge the idea of the constancy of composition of chemical compounds, implicit in the work of the vast majority of chemists including his influential colleague Lavoisier. Adducing many examples of variable composition systems, like glasses, alloys, and solutions, he suggested that most chemical systems are in fact equilibrium systems and that constancy of composition was an artefact of isolation conditions. This particular aspect of the Essay's claims led to a long-running controversy in the chemical literature with another French chemist, Joseph Louis Proust, that ended in victory for Proust and general acceptance of constancy of composition.

It is likely that this led to neglect of the majority of Berthollet's ideas in the Essay. These are about the nature of affinity; the factors that affect it; and the importance of equilibrium in considering chemical reactions. It is not far-fetched to claim that Berthollet's "Essai de Statique Chimique" is the first discussion of the field that nearly a century later would be termed Physical Chemistry.

CHAPTER 12: Berzelius "Jahres-Berichte".

Jons Jacob Berzelius (1779 – 1848) was a giant of early 19th. Century chemistry. He was informally known as the lawgiver of chemistry. To this day we owe so much of the everyday language and ideas of chemistry to this Swedish chemist. He coined such terms as isomer, polymer, catalysis etc. He turned the awkward quasi-alchemical symbols of even as enlightened a scientist as Dalton into the alphabetical symbols we use today: H, O, N, C, Fe; and so on. He standardized atomic weights. Perhaps even more important than these was his electrochemical view of chemical affinity that gave rise to the terms electronegative and electropositive. I felt I had to include a work by Berzelius among the "Great Chemistry Books". But which to choose? Berzelius was a prolific writer and his "Textbook of Chemistry", 6 volumes, 1808 – 1830, and written in Swedish, seemed at first the right choice. It was translated into German, French, Dutch, Italian, and Spanish – but not, curiously, into English. Clearly this was a widely read and influential text. But I have rejected that significant work in favor of a series of volumes that marked a new and enormously influential departure in the literature of chemistry.

Berzelius and elements he discovered.

Beginning in 1822 Berzelius published in Swedish the first of an annual and critical review of new work in chemistry, physics, and mineralogy. It was immediately translated into German and it is by its German title that it is generally known: "Jahres-Bericht uber die Fortschritte der physicalischen Wissenschaften" or "Jahres-Bericht" for short. In all 27 volumes were published over the period 1822 – 1848 (the year of Berzelius' death). A French version was published but only covered the years 1841 – 1848.

As far as I have been able to ascertain this impressive series of annual critical abstracts was the work of Berzelius alone. He read every important article or book on the physical sciences published in a given year; abstracted those he deemed important; wrote critical reviews of them; and arranged them appropriately for publication. My mind boggles at the thought that

Berzelius, having done all this, still had to proof the volumes before they were published. The team-written "Chemical Abstracts" owes everything to the one-man "Jahres-Bericht". Of course the latter was the parent of every other scientific abstract publication that followed it.

If you think that "Jahres-Bericht" was Berzelius' full-time job you are mistaken. He wrote books on chemical mineralogy; and animal chemistry (the biochemistry of animal systems). He was a prolific letter-writer – an important way for scientists to communicate in those days when journals were in their infancy. And he was a superb experimental chemist. One of his main objects was to strengthen Dalton's atomic theory by determining precise values of atomic weights. In that process, working on mineral samples, he discovered cerium (as its oxide, ceria) in 1803; selenium in 1817; and thorium in 1828. A little after Davy he isolated, by electrochemistry, potassium and sodium. He used a mercury electrode to prepare amalgams of the alkaline earth metals, and then produced the free metals. For those of us who have an interest in forensic science, Berzelius improved the Marsh test for arsenic (a favorite poison of the 19th and perhaps earlier centuries).

CHAPTER 13: Dalton: " A New System of Chemical Philosophy".

Dalton's biography may be familiar, but is always worth repeating.. Son of a Quaker weaver and his wife, a woman of strong character, Dalton was born in 1766 in a small village in England's Lake District in the north west of the country. He attended a village school and was then patronized by Elihu Robinson, a Quaker with interests in science, who instructed him in mathematics. At the age of 12 Dalton began teaching at his former school. When he was 15 he was employed along with his brother at a school in a local town, Kendal, where he became acquainted with the blind scientist John Gough who, among other instruction, led Dalton to the study of weather (the Lake District has a lot of weather!) that he maintained throughout his life.

In 1793, on the recommendation of Gough, Dalton was appointed as a teacher of mathematics and science at New College, Manchester, a school for children of dissenters, that is Christians who did not choose to belong to the Established Anglican Church. He remained on the staff of New College for the rest of his life. He also became a member, and later Secretary, of the Manchester Literary and Philosophical Society, and many of his publications appeared in its Journal. For the full account of how Dalton came to develop the atomic theory that bears his name I refer you to "Creations of Fire" by Cathy Cobb and Harold Goldwhite - a lively account of the history of chemistry.

The book "A New System of Chemical Philosophy" by John Dalton was published in 1808. It may be credited with the rapid development of chemical knowledge in the early 19th. Century. It is quite a slim volume but it includes details of most of Dalton's influential discoveries. It begins with a discussion of heat or caloric (a term widely used at the time). This chapter includes sections on specific heats, and heats of combustion, A curious discussion of the "Natural Zero of Temperature", or what we term absolute zero, concludes, on very shaky grounds, with a value of greater than 1500 degrees Fahrenheit below room temperature.

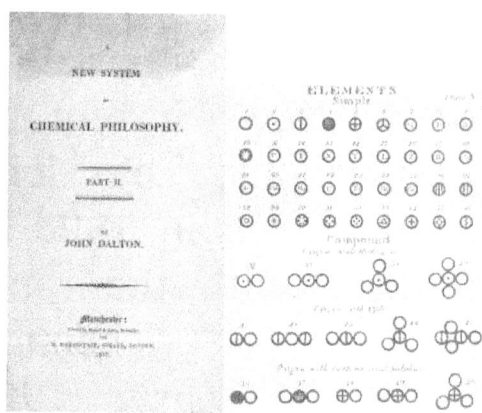

Dalton's "NEW SYSTEM" and his symbolism for atoms and compounds

The second section is about the constitution of bodies including gases, liquids, and solids. It includes an extended discussion of the development of Dalton's Law of Partial Pressures and some arguments contradicting what we now term Gay Lussac's Law of combination of gases.

The final, and very short, section finally gets to the heart of the matter: the atomic theory. It is misleadingly titled "On Chemical Synthesis". It includes the conservation of matter principle; a table of relative weights of 14 different atoms and 23 compounds; a set of diagrams, in Dalton's somewhat cumbersome pictorialism, of those materials; and Dalton's arbitrary but important principle of simplicity, that very principle that fixed water as HO, ammonia as NH, and led to 50 years of argument about atomic weights versus equivalents that finally yielded to the arguments presented in the next Great Book (spoiler alert: something of a cheat since it was only a pamphlet) to be discussed in the next chapter.

CHAPTER 14: Cannizzaro: "Sketch of a Course in Chemical Philosophy".

This chapter is about a very slim book, no more than a pamphlet, as I warned you in the last chapter, but one of the most influential publications in the whole of chemical history. It is Stanislao Cannizzaro's 1858 "Sketch of a Course in Chemical Philosophy". My personal copy is No. 18 of the valuable Alembic Club Reprints, an English translation of Cannizzaro's Italian original, published in Edinburgh, Scotland, in 1911 and reprinted in 1947.

ATOMI E MOLECOLE	SIMBOLI delle molecole dei corpi semplici, o formule fatte con questi simboli.		SIMBOLI degli atomi dei corpi semplici e formule delle molecole fatte con questi simboli.	
mo dell'idrogeno	H_2^1	=	H	=
cola dell'idrogeno	H	=	H^2	=
mo dell'ossigeno.	$O_2^1 = O_{15}^1$	=	O	=
cola dell'ossigeno ordinario	O	=	O^2	=
cola dell'ossigeno elettrizza- (ozono).	O_2	=	O^3	=
mo del solfo	$S_2^1 = Sn_2^1$	=	S	=
cola del solfo sopra 1000° binata).	S	=	S^2	=
cola del solfo sotto 1000° . .	Sn	=	S^6	=
cola dell'acqua	$HO_2^1 = HO_{15}^1$	=	H^2O	=
cola dell'idrogeno solforato	$HS_2^1 = HSn_2^1$	=	H^2S	=

Cannizzaro – and a Table from his "Sketch…"

In the first few decades of the 19th. Century, after Dalton announced his atomic theory, there was much confusion about how to go about In

In the early years of the 19th. century, after Dalton's atomic theory had been generally accepted, the problem of determining good atomic weight values remained. The confusion was so great that many important chemists, like Humphry Davy, abandoned atomic weights for equivalent weights (the mass of an element that combined with, e.g., exactly 8g of oxygen), claiming the experimental superiority of equivalents over the "theoretical" atomic weights. Cannizzaro, planning a course in general chemistry at the Royal University of Genoa, pointed the way forward, and summarized his approach, and his course, in this book.

I quote from the opening of the book:" I believe that the progress of science made in these last years has confirmed the hypothesis of Avogadro, of Ampere, and of Dumas on the similar constitution of substances in the gaseous state; that is, that equal volumes of these substances, whether simple or compound, contain an equal number of molecules..." Cannizzaro then goes on to outline how he develops this theme to arrive at unambiguous atomic weights for those elements that form gaseous compounds.

In a clear and logical presentation he cites the work of Gay-Lussac that so influenced Avogadro; examines the arguments of Berzelius' dualistic theory that led that great chemist to reject formulas like H_2 and O_2 for the molecules of these elemental gases, formulas that inevitably proceeded from the acceptance of Avogadro's hypothesis. Gas densities then become the key to Cannizzaro's determinations of the relative weights of a variety of molecules even before their formulas are known. He then defines atomic weight empirically: "The different quantities of the same element contained in different molecules are all whole multiples of one and the same quantity, which, always being entire, has the right to be called an atom." Cannizzaro applies this principle to hydrogen, oxygen, chlorine, and, perhaps most significantly, carbon. The volatile compounds of mercury are also examined and, by applying the specific heat principle of Petit and Dulong (not acknowledged by Cannizzaro in this work!) he derives atomic weights for a number of metals. Altogether a spectacular advance in chemistry.

Cannizzaro was one of the attendees at the first international chemistry conference, held at Karlsruhe in 1860. The agenda of the conference included trying to reach agreement on such fundamental terms as atom, molecule, atomic and molecular weights etc. The planning committee could not even agree on an agenda! Meanwhile Cannizzaro made his presentation and distributed copies of his pamphlet, most of which probably made it to various round files. However two attendees kept their copies and actually read them and later alluded to the powerful impression it made on them. Both were young university faculty charged with planning a beginning chemistry course. Both used Cannizzaro's work in planning their courses. And both kept thinking about and extending Cannizzaro's ideas. Their names were Victor Meyer and Dmitri Mendeleev. Each independently and almost simultaneously invented the periodic table. And the rest, as they say, is chemical history.

CHAPTER15: Faraday "Experimental Researches in Electricity".

Faraday is one of my heroes of science. If you wonder why, please read a biography of Michael Faraday (there is a short one in my book "A Chemical Chrestomathy" available on line). Born into very humble circumstanccs he taught himself chemistry and physics and rose to become an outstanding scientist. He may be remembered mostly for his discoveries in current electricity including the inventions of the electric generator, the electric motor, and the transformer among others, but he began his scientific career as a chemist, and it is his work in electrochemistry that forms a substantial part of this chapter's Great Book.

Faraday in middle age – a photograph.
We are now in the era of photography.

The "Experimental Researches" is a one volume compilation of the three volumes Faraday published under this title, and each volume is a compilation of the individual papers that he published in "Philosophical Transactions", the journal of the Royal Society. The papers span the period 1839 to 1847. I will focus my examination on the work in electrochemistry from the middle of that period.

Faraday set electrochemistry on a firm scientific foundation. He established the equivalence in electrochemical effects between electricity generated by a "static" electrical machine, such as a Wimshurst machine, and by a Voltaic battery. He determined the basic law of electrochemistry: the amount of electrochemical product depends on the current passed and the elapsed time; and is independent of potential difference, or voltage. Most significantly for further theoretical developments he showed that electrochemical and chemical equivalents are identical, suggesting that chemical change has an electrical origin.

Together with William Whewell he devised the nomenclature of electrochemistry. The terms ion, electrode, electrolyte, anode, cathode, anion, cation etc. are all Faraday and Whewell coinings. And now I will challenge my readers (as I do my students when I teach a course on History of Chemistry). What is the etymology of the an- and cat- prefixes for these terms for ions and electrodes ?

Faraday made many other contributions to chemistry that are not in this book and so are "outside scope". To mention a few: the first syntheses of benzene and some chlorinated alkenes; the first production of liquid ammonia and sulfur dioxide; and the observation that air becomes a better conductor of electricity as its pressure is reduced (the last two with Humphry Davy).

Faraday giving a Christmas lecture at the Royal Institution. Faraday, following Davy's example, was a popular lecturer- in this case with a Royal audience. Prince Albert and some of his children are in the front row.

To get a taste of Faraday's style read "The Chemical History of a Candle

INTERLUDE : The Twentieth Century

So far my discussions of great chemistry books have been restricted to one or two examples per time period. Perhaps the nineteenth century has been unfairly represented by these choices, but when it comes to the twentieth century it would be absurd to restrict the discussion to only a couple of books. More was accomplished in chemistry in the twentieth century than in the whole of the previous history of the subject. Many of the advances in chemistry were due to advances in technique. Where would we chemists be without nuclear magnetic resonance? Consequently my discussions of great twentieth century chemistry books will include not just one or two examples, but rather many examples. The choices are idiosyncratic; many come from personal experience. I hope you enjoy them – and feel free to send me your selections of great chemistry books at hgoldwh@calstatela.edu.

CHAPTER 16: Werner: "New Ideas on Inorganic Chemistry.

I begin my discussion of great 20th. century chemistry books with a work that was truly ground-breaking and that led to the development of inorganic and bio-inorganic chemistry in the 20th. and 21st. centuries. I refer to Alfred Werner's book, published in 1905, that I will refer to by an Englished version of its title as "New Ideas on Inorganic Chemistry".

Werner (Nobel Laureate in Chemistry in 1913) certainly did have new ideas. Born in Alsace he was educated in Switzerland, received his Doctorate in chemistry from Zurich, worked in organic chemistry on isomerism in oximes with Hantzsch in Paris, and then returned to Switzerland for the rest of his rather short professional life.

The problem of structure in the class of inorganic compounds we now call coordination compounds challenged chemists in the 19th. century. Influenced by the great successes of chain and ring structures in the organic realm they tried to develop similar theories to account, for example for the cobaltammines. Werner, using physical methods including conductivity data, saw the solution. He distinguished between the valence of the metal atom, that we now call its oxidation state, and the number of groups it could attach, now called the coordination number.

Werner also suggested that particular coordination numbers were associated with a particular geometry. The commonest coordination number, six, led to an octahedral arrangement of groups – ligands – around the central metal ion. This explained the existence of isomers of certain coordination compounds and led Werner to the prediction of optical activity in some compounds – a prediction he verified by synthesis and resolution of those compounds.

The structural theory proposed by Werner clarified the inorganic chemistry of coordination compounds in the same way as the structural theory of Kekulé and Couper had done for organic structures decades earlier.

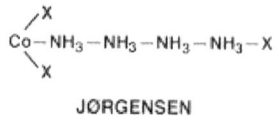

JØRGENSEN

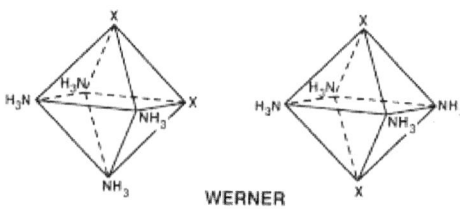

WERNER

Top: Differing structures for cobaltammines proposed by Jorgensen and Werner.

Below: Alfred Werner

CHAPTER 17: G.N. Lewis "Valence and the Structure of Atoms and Molecules".

I place next in my great chemistry books of the twentieth century "Valence and The Structure of Atoms and Molecules" by Gilbert Newton Lewis, published in the ACS monograph series in 1923. Lewis received his Ph.D. from Harvard, spent time in Europe and Asia, and was recruited to help build the Chemistry Department at U.C. Berkeley in 1912. He spent his career there, eventually becoming Dean of the College of Chemistry. His work in chemistry covered a wide range: thermodynamics including a seminal text; isotope separation; acid-base theory – we all know about Lewis acids and bases; the photon, so named by Lewis; and, of course, the subject matter of this Great Book. Although he helped teach many Nobel prize winners (Urey, Giauque, Seaborg, Libby, and Calvin) Lewis never won a Nobel Prize.

The Preface to this book begins with these words: " I take it that a monograph of this sort belongs to the ephemeral literature of science." He was wrong! This monograph is a classic of science. Its chapters proceed from the atomic theory and the Periodic Law to the chemist's picture of the atom including Lewis' own octet theory. Most of the rest of the book covers the union of atoms, that he calls the modern dualistic theory echoing Berzelius' dualistic theory of the early 19th. century. Lewis' new theory of valence is a key concept applied to molecular structure; and to covalent bonds including multiple bonds. He discusses exceptions to the octet rule and develops a magnetochemical theory of chemical affinity. And all in about 170 pages.

Lewis had come a long way in 1923 from the initial sketches that he made in 1908 of a primitive version of the octet theory. To this day, a century later, chemists the world over work with and discuss Lewis structures, and Lewis acids and bases.

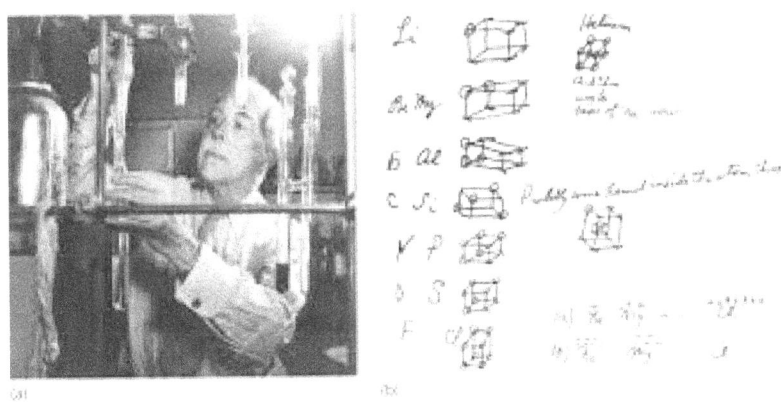

Lewis in the laboratory.

A 1908 sketch by Lewis of the octet rule.

CHAPTER 18: Louis P. Hammett: Physical Organic Chemistry: Reaction Rates, Equilibria, and Mechanisms" .

You can look forward to an examination of some physical organic chemistry texts in these chapters on "Great Chemistry Books of the Twentieth Century" and I present the first of these: "Physical Organic Chemistry: Reaction Rates, Equilibria, and Mechanisms" by Louis P. Hammett, Professor of Chemistry at Columbia University, published by Mc-Graw-Hill in 1940. Hammett (1894 – 1987) was educated at Columbia and was a pioneer investigator in physical organic chemistry. He also worked on superacids and devised a function to quantify their acidity known as the Hammett function. But he is best known for the Hammett equation relating rates and equilibria in organic reactions; a function that is explored in depth in his text. His many awards and recognitions included the Priestley Medal of the ACS; the Willard Gibbs Award; the National Medal of Science; and the Barnard Medal.

Now to the text. My copy is the Fifth Impression of the First Edition, indicating its appeal to chemists of the 1940s. I quote from the Preface: " in the last two decades on the borderline between physical chemistry and organic chemistry....there has grown up a body of fact, generalization, and theory that may properly be called "physical organic chemistry.....the study by quantitative methods of the mechanism of reactions and of the related problem of the effect of structure and environment on reactivity".

Hammett's Physical Organic Chemistry" is not a tome; after all the subject, while not in its infancy, was still in its childhood. It measures just under 400 small pages. The first two chapters discuss quantum views of structures of non-electrolytes and electrolytes. They sketch, rather than explore in depth, such topics as atomic structures, valence bond theory, and Pauling's resonance theory. At that time hydrogen-bonding was a challenge to current theories.

Chapter III, on equilibrium and energy of reaction, explores the significance of Boltzmann's equations in understanding the connection between these parameters. Chapter IV, on reaction rates and mechanisms, seems to me to be a central part of this text, and develops the basis of what later became known as the Hammett equation – perhaps you encountered this in your organic chemistry courses. Another key to Hammett's original thinking is in Chapter VII, 'The Effect of Structure on Reactivity" that explores quantitative relationships between systems of reactions, looking, for example, at correlations between hydrolysis rates of esters and the ionization constants of their respective acids. This chapter includes the explicit statement of Hammett's equation.

The Hammett Equation

Louis P. Hammett (1894-1987)

Hammett ranges far and wide in his exploration of organic reactions and includes displacement reactions and their stereochemistry; enolization; carbonium ions; carbonyl additions; and atomic, radical, and redox reactions. I believe that this 80-year-old classic text, surely one of the great books of chemistry in the 20th. century, still has something to teach chemists of the 21st. century.

CHAPTER 19: Linus Pauling: "The Nature of the Chemical bond: and the structure of molecules and crystals".

"The Nature of the Chemical Bond: and the structure of molecules and crystals" by Linus Pauling was first published in 1939 and was based on a series of Baker Lectures given by Pauling at Cornell University in 1937-38. (Incidentally this lecture series has produced a number of Great Chemistry Books that I will be commenting on in later chapters.) My own copy of this important book is the third edition, considerably revised and updated, published in 1960, and dedicated to Gilbert Newton Lewis.

This third edition is quite inclusive and in its over 600 pages uses extensively Pauling's important contribution to bonding theory, namely the concept of resonance. The first section of the book is titled "Resonance and the Chemical Bond". In this section Pauling describes formally, in terms of wave functions, the concept of an actual molecular structure being, in certain cases, a resonance hybrid of alternative structures. He applies the idea of resonance in this section to the simplest of all molecules, the hydrogen molecule-ion H_2^+. This is just for starters!

Pauling's scope in this book is very wide. He moves from the electronic structures of atoms and the rules for the formation of covalent bonds to electronegativity and the "Pauling Scale" and the partially ionic character of covalent bonds. Commentaries on bond angles are followed by bonding in coordination compounds and a brief discussion of ligand field theory.

Two chapters give accounts of resonance of molecules among different valence bond structures; and different types of resonance in simple molecules and ions. Sandwiched between these chapters is an introduction to interatomic distances and structures of molecules and crystals. In a subsequent chapter Pauling discusses one-electron and three-electron bonds and electron-deficient compounds including the boranes.

Pauling at the blackboard discussing protein structures – work that came well after his book on "Valence …"
Pauling won Nobel prizes for both Chemistry and Peace.

One of the most interesting sections for me is the chapter on the metallic bond, and I can't resist including a couple of personal reminiscences here. I was an undergraduate at Cambridge University in the early 1950s and we had a short course in our inorganic chemistry class on metallic structure, illustrated by bubble tray models. The course was given by Professor Lawrence Bragg, then at the Royal Institution in London. Yes, the Bragg of Bragg's Law, who along with his father won the Nobel Prize in physics in 1915 when he was 25 years old. In the 1980s I had the pleasure of attending a lecture by Pauling at Cal. State, Los Angeles, in the Leon Pape lecture series, and he too discussed metallic structures.

CHAPTER 20: Weeks and Leicester: Discovery of the Elements.

In my continuing pursuit of the 20th. Century's great chemistry books I have re-encountered a work of chemical history that I consider a "Great Book". It is "Discovery of the Elements" by Mary Elvira Weeks and my 7th. Edition, published by the Journal of Chemical Education in 1968, is co-authored by Henry M. Leicester, who claims this edition to be completely revised with new material added. This generously illustrated volume runs to nearly 900 pages and is, as its title suggests, a detailed and accurate account of the ways in which the elements known at the time of publication, 104 in number, were discovered.

The mother of this enterprise, Mary Elvira Weeks, (1892 – 1975) was the first woman to be awarded a Ph.D. in chemistry by The University of Kansas, and the first woman faculty member in that department after spending 7 years as a high-school teacher. Her doctoral thesis was on the effects of pH in precipitation of calcium and magnesium salts. She spent 22 years at Kansas, mostly as a teacher, and from 1932 to 1933 she published more than 20 articles in the Journal of Chemical Education that became the nucleus of the first edition of "Discovery of the Elements" published in 1933. The book was well reviewed and five more editions followed. In 1946 Weeks collaborated with Charles E. Browne in writing "A History of the American Chemical Society – Seventy-five Eventful Years" which appeared in 1952, five years after Browne's death. In 1967

Weeks was awarded the Dexter Prize by the Division of the History of Chemistry of the American Chemical Society.

Mary Elvira Weeks

The 7th. Edition of "Discovery…" was prepared by another distinguished historian of chemistry, Henry M.

Leicester. After earning a Ph.D. at Stanford in organic chemistry in 1930 he spent time in Europe and was a faculty member at a number of institutions before settling at the University of the Pacific where he was Professor of Biochemistry for 20 years. He developed expertise in the history of chemistry in Russia and published biographies of Russian chemists in many authoritative sources. He founded the series "Chymia" , an annual publication containing articles about the history of chemistry, and co-edited a historical source book in chemistry. He also was awarded the Dexter prize – in 1962.

The 7th. Edition of "Weeks and Leicester" is regarded as a classic in chemical history. Its in-depth approach can be gauged by examples: the chapter on three important gases, hydrogen, nitrogen, and oxygen, covers 63 pages. Three alkali metals, potassium, sodium, and lithium, occupy 45 pages. Every page has footnotes, usually several. There is a 16 -page chronology from the 16th. to the 20th. Century, and both name and subject indexes. The writing is fluent and draws the reader in.

A final note. It has been over 50 years since this history of the discovery of the elements was published and surely an update and an 8th. Edition is overdue. Other good books on the history of the elements have been published in that time, but none of them comes near the depth and authoritative approach of Weeks and Leicester. So this is another Challenge to my Readers.

CHAPTER 21: William Waters: "Physical Aspects of Organic Chemistry".

When I was in graduate school one of the leading topics of discussion was physical organic chemistry, an area that seems to have receded in importance somewhat at present. My interest in that subject was reinforced in my post-doctoral stay at Corncll by the visit of Sol Winstein from UCLA to give a Baker lecture course on physical organic chemistry. Winstein's tragically early death meant that the textbook that usually summarizes and expands a Baker lecture series did not appear. This reminiscence was prompted as I scanned my bookshelves for candidates for my "Great Chemistry Books : the 20th. Century" chapters. I came across a book I must have purchased years ago but had completely overlooked: "Physical Aspects of Organic Chemistry" Second Edition, by William A. Waters of the University of Durham, with an introduction by Professor Martin Lowry, published by Van Nostrand in the U.S.A. in 1937. The first edition was in 1935 - an early text on the subject and a very well written one. My copy of this book was released from the library of Juniata College, a private liberal arts college in Huntingdon, Pennsylvania. Founded in 1876 as a co-educational school, it was the first college started by members of the Church of the Brethren as a center for vocational learning for those who could not afford formal education.

In glancing over the Preface and Introduction it is clear that the original plan for this text was Lowry's, but it was Waters who completed it. Lowry acknowledges the help of Dr. C. P. Snow in preparing the more physical chapters at the beginning of the book. Some of you may be familiar with C. P. Snow as the author of many mid-20th. century best-selling novels. I recommend two of them, in particular, to my readers. "The Search" poses some challenging questions about originality, plagiarism, and falsification in the conduct of scientific research. "The Masters" is an enthralling examination of academic politics in the context of an Oxbridge college.

The opening paragraphs of the Waters and Lowry text have an immediate appeal to this historian of chemistry. Under the heading of "Berzelius's Dualistic Theory of Chemical Affinity", a theory based on electrical attraction and that gave us the terms electronegative and electropositive, the book leads us through the historical events that led to Berzelius's ideas – still a foundation of theories of the interactions of atoms and molecules. This historically designed approach leads logically to Dumas, van't Hoff and le Bel, Kossel and G.N.Lewis, and Alfred Werner – the revival of dualism as the authors put it.

After this historical and physical introduction the book goes on to unsaturation and free-radicals. This latter subject starts with Gomberg and the triphenylmethyl radical, the first isolated organic free-radical. This type of radical was characterized, among other measurements, by its paramagnetism. Nitrogen centered free radicals are produced by the partial dissociation of tetra-aryl hydrazines and oxygen centered free radicals similarly from peroxides.

Molecular rearrangements occupy a long chapter, that includes many examples and suggestions about mechanisms. A discussion of Walden inversion draws no conclusions about mechanism, while the pinacone-pinacoline and Beckman rearrangements fare no better. But recall- this is over 80 years ago. We have come a long way since.

Physical organic chemistry reached maturity from the 1940s on with the publication of a number of seminal texts that do deserve designation as "Great Books of Chemistry of the 20th. century".

CHAPTER 22: C.Coulson "Valence".

I come now to a book about a central theme in chemistry, a concept first perceived by Edward Frankland in the 19th. century. I refer to valence. The book under consideration is "Valence" originally published by Charles Coulson, Professor of Theoretical Chemistry at Oxford University, in 1952. I was an undergraduate at "the other place", namely Cambridge, in the 1950s, and Coulson's book was my bible as I wrestled with the concepts and mathematics of theoretical chemistry. "Valence" was substantially updated in a 3rd. edition by Roy McWeeny, a Coulson student, and Professor of Theoretical Chemistry at the University of Sheffield. I have mislaid or lost my copy of the first edition of "Valence" but I have to say that the third edition, which appeared in 1979, marks a substantial advance on the first edition and is a "Great Book" in its own right. Incidentally my paperback copy, published by Oxford University Press, cost me 8 pounds 50 pence – at a time when you could buy a decent fish and chips lunch for 5 shillings. You do the math!

Coulson in his Oxford Office

An initial chapter discusses the features that any theory of valence must account for, such as why do molecules form?; how do we account for their stereochemistry?; why do bond angles and bond lengths vary in a series of related molecules? etc. Chapters 2 and 3 plunge into the heart of modern theories of wave functions, atomic orbitals, and the principles of wave mechanics. Chapters on diatomic molecules follow, exploring wave functions for pairs of electrons – the familiar covalent bond is explained in detail – and examining such fundamental ideas as hybridization, electronegativity, and bond polarity.

Chapter 7 on polyatomic molecules discusses localized and non-localized orbitals and hybridization, extended to the involvement of d-orbitals in understanding the structures of such non-octet molecules as SF_6 and PCl_5. An extensive chapter on carbon compounds follows covering conjugation, benzene and aromaticity, Hückel theory, and effects of heteroatoms. Then the text branches out to transition metal compounds and covers both crystal field and ligand field theories; high and low spin complexes; and pi-bonding and sandwich molecules such as ferrocene.

The subject of chemical reactivity, after covering substitution reactions, goes on to examine in detail the well-known Woodward-Hoffman rules, and the more obscure Dewar-Evans rules relating to aromatic reactions. Electronic theories of the solid state cover energy band theory and metals and semiconductors; Brillouin zones; and crystal types. Then follows a chapter on weak interactions and unusual bonds. This includes hydrogen bonding; electron-deficient molecules such as the boranes; phosphazenes and related cyclic compounds; and noble-gas compounds.

In a final chapter the author covers the deficiencies in the then (1979) current theories and embarks on an exploration of ab-initio self-consistent field (SCF) theory. In a summarizing section he points to the advances in theoretical chemistry made possible by computation, but draws attention to the great costs of computation for complex systems. In the 2020s those costs have become miniscule, thanks to advances in computer technology, but even in the 2020s this book on valence would still be a valuable introduction to the subject. And its timelessness is what makes it worthy of inclusion in the "Great Chemistry Books" of the 20th. century.

CHAPTER 23: C.K.Ingold: "Structure and Mechanism in Organic Chemistry"

In an earlier chapter I mentioned a possible succession of physical organic chemistry texts that I might include in the "Great Chemistry Books : 20th. century". Here we go with another celebrated volume: "Structure and Mechanism in Organic Chemistry" by C.K. Ingold, Professor of Chemistry, University College, University of London, published in 1953. This is another of the seminal chemistry texts based on the remarkable series of Baker Lectures presented at Cornell University. Ingold visited Cornell in Fall Semester of the 1950 – 51 academic year to present this lecture series. The final product was this 800- page tome covering, as I shall detail, most mechanistic aspects of the chemistry of organic molecules undergoing homogeneous molecular reactions in their normal states.

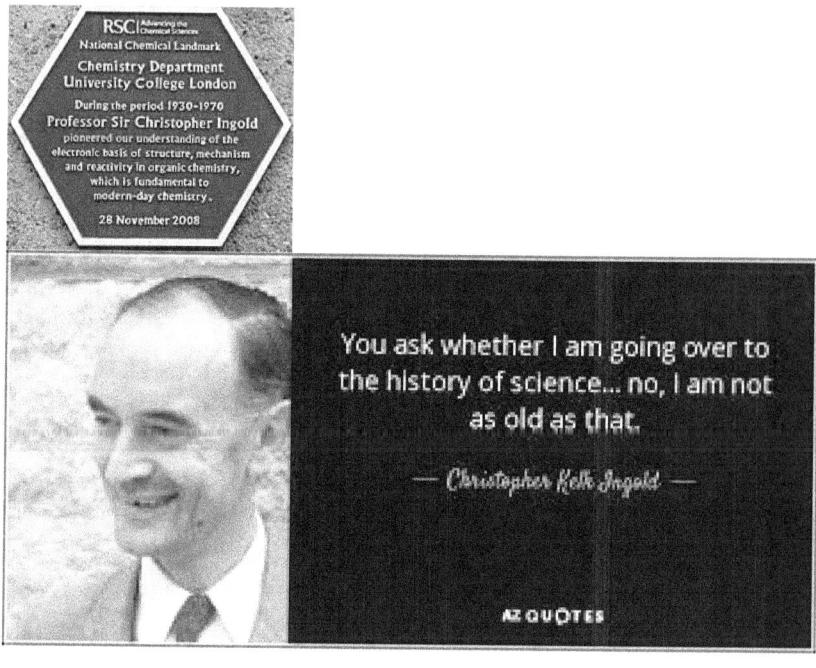

When I was in undergraduate organic chemistry classes in the early 1950s we heard very little about physical organic chemistry. However the names of Hughes and Ingold did come up in the context of nucleophilic aliphatic substitution, along with the Sn1 and Sn2 symbolism. The two collaborators had been working on the mechanisms of these classes of reactions since the 1930s and had developed the Sn taxonomy and a rationalization of mechanisms of these classes of reaction.

In Ingold's list of acknowledgments in the book's Preface there is a litany of mid-20th. century contributors to physical organic chemistry. In addition to Hughes there are J. F. Bunnett (Reed College); J. W. Baker (Leeds); J.D. Roberts (then at M.I.T.); and F. H. Westheimer (then at Chicago).

The first two chapters of Ingold's book give the background of valency and molecular structure; and interactions (forces) between and within molecules. This second chapter presents an in-depth examination of the relative strengths of the forces detailed, a subject that Ingold had previously reviewed in detail in the early 1930s.

Physical properties at that early date were mostly based on macroscopic determinations of such parameters as electric dipole moments; heats of formation and reaction; electrical polarizability; and bond force constants from spectroscopy. An exploration of aromatic character moves from theory to physical properties with a detailed exploration of diamagnetic polarizability in unsaturated molecules, conjugated unsaturates, and aromatics. It is impressive to realize how far chemists of nearly a century ago could go in understanding molecular properties based on a handful of bulk-matter properties.

Two chapters of about 100 pages each present what I characterize as the heart of the matter. The first on electrophilic aromatic substitution draws together masterfully the experimental evidence on this subject looking at electronic effects on the ratio of ortho- to para-substitution, and at steric effects in both the reagents and the substrates. Nitration serves as a model for this class of reaction, shedding light on halogenation, acylation, sulfonation, mercuration, hydrogen exchange (drawing on isotopic labelling experiments carried out in the early 1930s !), and diazo-coupling.

Nucleophilic substitution at saturated carbon atoms had been a long-term study of the Ingold and Hughes group, and its treatment in this text is complete and comprehensive. Polar effects, solvent effects, salt effects, and catalysis are all scrutinized and interpreted. The Walden inversion gets its due. And the effects of steric hindrance on both kinetics and thermodynamics of nucleophilic substitution are explored.

We are still less than half-way through this comprehensive volume and remaining topics include eliminations; rearrangements; additions to olefins; acids and bases; carboxyl reactions; and nucleophilic aromatic substitution. A ten page index pays tribute to the thoroughness with which Christopher Ingold prepared the book. Almost every page is footnoted to the original sources.

Ingold (1893 – 1970) received many honors. Fellow of the Royal Society in

1926, he was knighted in 1958 for services to British scientific education.

CHAPTER 24: John. E. Leffler of Florida State University, and Ernest Grunwald: "Rates and Equilibria of Organic Reactions".

Continuing my very personal review of the Great Chemistry Books of the 20th. century I come to yet another physical organic chemistry text. My choice for this chapter is "Rates and Equilibria of Organic Reactions" by John. E. Leffler of Florida State University, and Ernest Grunwald of Bell Telephone Laboratories Inc. This book was published by Wiley in 1963. It seems that the first seven decades of the 20th. century was something of a golden age for physical organic chemistry.

The subtitle of the work indicates that the treatment will be by statistical, thermodynamic, and extrathermodynamic methods and it begins with a remarkable Apologia quoted from the philosopher Bertrand Russell: "Unless we can know something without knowing everything, it is obvious that we can never know something." Clearly Leffler and Grunwald believe we can know something. In their preface they outline their approach to the subject and I quote: "Organic reactions usually involve complex molecules reacting in the liquid phase. The theoretical analysis of their rates and equilibria is necessarily less exact than that of small molecules reacting in the gas phase…the starting point is a macroscopic or thermodynamic analysis to which is added just enough microscopic detail to allow explicitly for structural and medium effects." And the text delivers on this promise in 450 pages.

After a 6- page glossary of symbols (!) the opening chapter dives right into equilibrium from a statistical point of view, that is equilibrium as an exercise in probability. Naturally this includes applying Boltzmann distributions to chemical systems starting with a consideration of the configurations of 1,2-dichloroethane. This leads to a discussion of partition functions for various systems. It is clear right from the start that Leffler and Grunwald take the physical aspects of physical organic chemistry much more in depth than either Hammett or Ingold do in previous texts I have discussed in this series.

In addition to expected chapter headings including "Equilibrium and the Gibbs Free Energy", "Free Energy, Enthalpy, and Entropy", "Concerning Rates of Reaction", and "Rates of Interconversion of Subspecies", all treated in physical terms and occupying the first third of the book, the authors then embark on extrathermodynamic relationships.

Chapter 6 is a theoretical introduction to this concept. Not the least of the pleasures in reading this book are the epigraphs that adorn each chapter. That for Chapter 6 is by Havelock Ellis and begins: "For even the most sober scientific investigations in science, the most thoroughgoing positivist cannot dispense with fiction…"

The authors then go on to examine the fictions, or simplifying models, that are essential in trying to explain trends in reactivity in organic reactions. An example is trying to separate substituent effects and medium (solvent) effects. They consider such reactions as rearrangements of aminotriazoles, hydrolyses of esters catalyzed by imidazoles, and epoxidation of stilbenes. Another aspect of this modeling process is the existence of simple additivity rules in explaining properties of molecules containing several substituents.

The extrathermodynamic interpretations go on to examine, in two chapters, free-energy relationships, considering first substituent effects and then medium effects. A concluding chapter on extrathermodynamic effects analyses enthalpy and entropy changes.

The last chapter in this stimulating book is unusual in discussing mechanochemical effects, a subject that is currently receiving much attention. The authors review reports of these effects in liquid phase reactions including effects of viscosity changes in mechanical polymer degradation. An example of a reaction in a crystalline matrix is the transformation of phenolphthalein from colorless to red upon the application of either pressure or shearing forces. They also discuss thermodynamics and reactions of fibrous polymers.

I do not know (perhaps you do) of a text that explores in a more rigorous way the physical aspects of physical organic chemistry, and that is why Leffler and Grunwald belongs in the canon of Great Chemistry Books.

CHAPTER 25: Primo Levi: "The Periodic Table".

When, a little while ago, I was discussing my development of a work on great chemistry books with a good friend, who happens to be an Emerita Professor of English, and who is married to a chemist, she suggested that I include this book. At first I was hesitant. Is this a great chemistry book in the usual sense? Perhaps not – but it is a great book, and its framework is the periodic table, emblematic of chemistry. I have decided to include it, and of course you can skip this chapter if you choose!

Primo Levi was born in 1919 in Turin, Italy, into a Jewish family. He graduated with honors from the University of Turin in 1940 but his chemistry degree diploma was stamped "Of Jewish Race" and, unable to find a job, he procured false papers and was employed by first a mining company and then by a pharmaceutical firm in Milan. His father's declining health drew him back to Turin in 1942. He then found that his mother and sister were in hiding in the country to avoid the increasing persecution of Jews in Italy. After his father's death in 1942 the surviving Levis left for the north and Primo joined the resistance. He and some comrades were arrested in late 1943 and on Primo's disclosure that he was Jewish he was sent first to an Italian camp and then in February 1944 to Auschwitz. He became number 174517; the tattoo remained permanently on his forearm. Because he was a chemist he was assigned to work in a synthetic rubber factory and escaped some of the worst horrors of Auschwitz – but he witnessed them. Liberated in January 1945 by the Red

Army Levi was one of the fewer than 10% of the over 7,000 Italian Jews deported to concentration camps to survive.

After his liberation, while working as a paint chemist, Levi wrote "Survival in Auschwitz", also titled "If this is a Man" as his eye-witness account of his camp experiences. Published in 1947 it initially received little attention.

"The Periodic Table" was published in Italian in 1975. Each of its 21 chapters is an individual short story and bears the name of a chemical element starting with argon an ending with carbon. The short stories are mostly autobiographical and connected in some way with each of the title elements. For example in Chapter 2, Hydrogen, Primo and a friend experiment with simple electrolysis of water. Some of the chapters may be fictional short stories.

In 1977 Levi became a full-time author and published several more novels and memoirs. "The Periodic Table" was published in English in 1984 and was hailed by famous writers and critics as an outstanding book. In 2006 a reviewer writing in the journal of The Royal Institution of Great Britain called it the best science book ever. And that is one reason why

I include it among the great chemistry books.

CHAPTER 26: Oliver Sacks: "Uncle Tungsten: Memories of a Chemical Boyhood".

This book is a personal choice of a warm and engaging memoir about how a young person became passionately interested in science, starting with chemistry. Personal because so many episodes in the book mirror my own experiences in coming to love chemistry.

Oliver Wolf Sacks was born in 1933 in London England. We are almost contemporaries; I am eighteen months older. After boarding school (to escape London's bombing) he attended Queen's College Oxford (I attended Queens' College Cambridge – note the different placement of the apostrophes), turned his interests towards medicine, and became a distinguished and controversial neurologist.

His book "Awakenings" published in 1973 dealt with treatment of certain types of Parkinson's disease. It was adapted into a successful movie with the same title. Many more books followed. The most successful was "The Man Who Mistook His Wife for a Hat" published in 1986.

"Uncle Tungsten" was published in 2001. The title refers to one of Sacks's uncles, who was employed by a maker of incandescent light bulbs that had tungsten filaments. He showed young Oliver a chunk of the extremely dense metal that the Uncle described as the metal of the future. Oliver developed an early interest in chemistry and more generally in science that led him to his chosen profession. "Uncle Tungsten" is a great book to give to any young person who might be showing   inclinations   towards pursuing a career in science.

CHAPTER 27: The final Chapter: Saul Winstein "Physical Organic Chemistry" (The Great Book that Never Was).

This is another very personal chapter and I must include some of my own history. While I was working on my Ph. D. at Cambridge in early 1956 Professor Emeléus, who had given a Baker Lecture series at Cornell the year before, received a request from Professor William T. Miller of Cornell to recommend a graduate student for post-doctoral work. I applied and was accepted and made my first trip to the U.S. in August 1956. I had a productive and instructive 2 years at Cornell and, not incidentally, met and married my wife there. Marie was doing academic programming in the Cornell Computing Center on a vacuum tube IBM machine with about

30K memory, wired programs, and punch card output. It occupied a large room and was hot in winter when the snow outside was a meter deep! My wife was a pioneering female programmer at a time when the job didn't even have a name.

But to return to our subject. The Baker lecturer for Spring 1957 was Saul Winstein from UCLA. He would talk on the topics of " Neighboring Groups, Solvolysis, and Rearrangements in Organic Chemistry", a topic of great interest to Professor Miller and his group. A major area of investigation for our group was the Sn2' rearrangement in the attack of nucleophiles on unsaturated highly fluorinated allylic compounds.

The George Fisher Baker Non-Resident Lectureship in Chemistry was established by endowment in

1926. There were usually 2 series a year, and it was expected that a book including the material of the lectures would be published in due course. Some notable lecturers include in 1926 Paneth on Radio Elements as Indicators; 1927 Walden on Salts; Acids and Bases; and Stereochemistry; 1929 G. P. Thomson on Wave Mechanics of Free Electrons; 1932 Stock on Hydrides of Boron and Silicon; 1933 Hahn on Applied Radiochemistry; 1937 Pauling on The Nature of the Chemical Bond; 1939 Debye on Determination of Molecular Structure; 1948 Flory on Principles of Polymer Chemistry; 1950 Ingold on Structure and Mechanism in Organic Chemistry; and 1953 Steacie on Photochemical and Free Radical Reactions. Several of the Baker books have been the subjects of earlier Chapters.

Baker Laboratories, Cornell University, Ithaca N.Y.

Saul Winstein (1912 – 1969) was born in Montreal and came to the U.S. at the age of 12. He attended UCLA and graduated with a B.S. in chemistry in 1934 with highest honors. Glenn Seaborg was a classmate. Winstein stayed on at UCLA to complete an MA with W. G. Young on allylic rearrangements. He then moved across town to Cal. Tech. working with Howard J. Lucas for a Ph.D. on hydration of alkenes; silver – alkene complexes; and substitution reactions of butenes. After an NRC fellowship at Harvard with Paul Bartlett from

1940-41 he was an Instructor at I.I.T. before being hired by UCLA in 1941. He was promoted to Professor in 1947. During W.W.II he collaborated with T.L.Jacobs on synthesis of antimalarial drugs. By 1948 Winstein had authored over 50 articles in JACS and J.Org. Chem. He received the ACS Award for Pure Chemistry for work on the role of neighboring groups in replacement reactions. He was made a member of the National Academy of Sciences in 1955.

I attended all of Winstein's Baker lectures, collected his 32 pages of mimeographed handouts, made notes (64 pages), and together with Professor Miller met with him several times to discuss our research project. Unfortunately in the confusion of numerous moves of my office in the past 20 years my folder has been lost. What follows are the notes I made when it was still in my possession. I gave a talk on this subject to the ACS

Division of History of Chemistry in New Orleans in 2007.

The breakdown of the lectures was: 2 on Sn1 and Sn2 reactions; 3 on solvolysis; 4 on ion pairs and salt effects; 3 on neighboring group participation by heteroatoms; 6 on carbon neighboring groups; 2 on hydrogen as a neighboring group; 2 on ring size and rates; 1 on complex rearrangements; and 1 on neighboring groups in addition reactions. In addition to citing numerous articles Winstein recommended as texts C.K.Ingold's book (Chapter 22); J. Hine "Physical Organic Chemistry", McGraw-Hill, 1956; M. Newman,ed. "Steric Effects in Organic Chemistry", Wiley, 1956; and two articles in Chem. Reviews, 1956: A. Streitweiser on solvolytic displacements at saturated carbon atoms; and R.H.DeWolfe and

W.G.Young on substitution and rearrangement reactions of allylic compounds.

What were the lectures like? Winstein was careful and detailed. He was an engaging lecturer who encouraged comment and interaction. He covered previous and contemporary work, including his own. I noted his comment on bridged ions: "carbonium ions may have bridged or non-classical structures and these may be encountered in connection with neighboring group participation".

A

B

Non-classical   structures   for
bridged ions.

It is clear, even from this truncated
summary, that there was ample
material in Winstein's Baker Lectures
for a substantial book. But no book
appeared. It is true that Winstein died
tragically young in 1969 but that was
12 years after the lectures. I speculate
that the appearance of the recent
(1956) texts and 1956 Chemical

Reviews articles covered much of what Winstein might have written on. The decade from 1957 – 67 was a very active one in physical organic chemistry and Winstein and his group were fully involved. This was the period of controversy over non-classical ions involving H. C. Brown and Winstein.

H. C. Brown and S. Winstein

In the UCLA archives there are 44 boxes of Winstein's papers. Frankly I no longer have the energy to explore this vast archive. Perhaps one day a more energetic and intrepid chemical historian will dive into this treasure trove and give us the definitive history of a great chemistry book that never was.

www.ingramcontent.com/pod-product-compliance
Lightning Source LLC
Chambersburg PA
CBHW070546220526
45467CB00003B/1089